Chemical Sensors for the
Detection of Mercury Toxicity

Chemical Sensors for the Detection of Mercury Toxicity

Shiva Prasad Kollur
Chandan Shivamallu

ELIVA PRESS

Published by Eliva Press
Email: info@elivapress.com
Website: www.elivapress.com

ISBN: 978-1-63648-371-9

© Eliva Press, 2021
© Shiva Prasad Kollur, Chandan Shivamallu
Cover Design: Eliva Press
Cover Image: Freepik Premium
Printed at: see last page

Table of contents

Chemical Sensors for the Detection of Mercury Toxicity

Shiva Prasad Kollur[1] and Chandan Shivamallu [2]

[1] Department of Sciences, Amrita School Arts and Sciences, Amrita Vishwa Vidyapeetham, Mysuru Campus, Mysuru – 570 026, Karnataka, India.

[2] Division of Biotechnology and Bioinformatics, School of Life Sciences, Department of Water and Health, JSS Academy of Higher Education and Research, Mysuru – 570 015, Karnataka, India.

Abstract

The harmful impact of mercury on biological systems is of great concern. Regardless of the efforts made by the regulating agencies, a decrease in Hg^{2+} concentration has not been a realized and hence mercury accumulation in the environment remains of utmost concern. Designing novel and efficient probes for recognition and detection of toxic metals in environmental samples has been of primary importance. Amongst the available techniques, probe designs involving the study of spectral properties has been preferred because of its obvious ease of instrumentation. Furthermore, occurrence of significant changes in the visible portion of electronic spectra enables detection by the naked eye. Thereby endorsing the preference for development of probes with off-on binary responses to aid in the in-field sample analysis. The prominence is further streamlined to the use of fluorescence to help characterize on-response the cellular detection of Hg^{2+} with ease. . In order to overcome the problem of developing efficient probes or sensors bearing fluorescence on-response mechanism that can work effectively in the physiological conditions various methodologies, such as chemo-dosimetric reaction mechanisms for the designing of new luminescent ligands, are being adopted. Additionally, modified charge transfer processes are also being considered for optical detection of the mercury (II) ion. In this review, all such possible techniques have been discussed in detail.

Keywords: Mercury toxicity; chemosensors; fluorescence; colorimetric sensors.

1. Introduction

Mercury, previously named Hydrargyrum (Hg) and commonly called quicksilver, is a chemical element with the atomic number 80 and a relatively high atomic mass of 200.59u. This element, which is also the only known liquid metal, is the second most toxic element on earth preceded by Plutonium. Mercury spike in water is relatively rare with the sources of the metal as common as the fossil fuels. With the expansive use in hydroelectric, mining, pulp and paper industries, the levels of mercury in the environmental has increased over time. Mercury is a common pollutant of drinking water and can cause kidney complications if present in amounts more than 2 ppb [1-3]. However, vaporization of liquid mercury leads to gaseous mercury, which is poisonous due to its nature of being absorbed into the blood [4]. With regard to the bioaccumulation prospect, mercury enters the top of the food chain via aquatic bacteria in the form of methylmercury, leading to a neurological condition, "Minamata disease"[5-7]. Elemental mercury is considered more lethal form and usually enters the system via the cutaneous and respiratory routes. Elemental mercury is known to cause cardiovascular, neural, and renal complications by inflicting DNA damage [9, 10]. Mercury methylation increases its lipophilic properties, posing a threat to the central nervous system [11]. Taken together, the high risk of toxicity associated with mercury exposure has incentivized global efforts towards potable water purification and mercury toxicity reduction. . Concerns about the deleterious effects of mercury poisoning have also motivated researchers to develop novel, affordable, and rapid detection tests that may be applicable to both the environmental and biological systems [12-14]. Therefore, a simple and convenient sensor that could solve the situation at hand rapidly, have sufficient sensitivity, and be resistant to interference by other metal ions is very timely. The developments so far have accounted for the expensive, sophisticated instruments, enabling complicated procedures or low sensitivity and selectivity approaches like atomic absorption, emission spectroscopy, or individually coupled plasma mass spectroscopy [15]. Despite efforts from the various regulatory agencies for reducing the mercury emission in the environment, the mercury contamination from natural and industrial processes across the globe has remained a serious threat to the human race in the last few decades. Therefore, monitoring mercury is of paramount importance to both environmental and human health.

In recent years, sensors for the recognition of heavy metal ions have been developed, which has attracted much attention from environmental scientists and chemists. Colorimetric receptors were developed for the selective recognition of heavy metal ions that received attention for a few decades due to their ability to permit naked eye detection of color change and simplify the whole method. However, chemosensors remained a reliable technique for the effective detection of mercury toxicity. Fluorescent chemosensors are preferred for a ratiometric response due to the ratio between two emission intensities that can help correcting the sensor analytic concentration, as well as environmental effects such as polarity, photo-bleaching and temperature. The expectations lead to the development of reversible ratiometric chemosensors capable of detecting Hg^{2+} [16]. Chemosensors with selectivity for specific targets of metal ions are continuously demanded. Those targeting the toxic heavy metal ions remain a vital category of chemosensors. Recognition of the detrimental effects of certain transition and post-transition metal ions on humans and animals has, in part, inspired work to develop compounds that selectively respond to specific metal ions for use as ion sensors. Research on fluorescence sensor capable of detecting the heavy and transition metal ions has gained attention due to significant progress in synthesis of novel fluorophores and the development of cost-effective yet efficient methods. Some advantages such as high sensitivity, selectivity, short response time, naked-eye detection, and fluorescence detection have been regarded the promising candidates in the molecular recognition and current applications of chemo-sensors. Furthermore, in biological and environmental systems, Hg^{2+} sensor interactions commonly occur as aqueous solutions. Hence, enough attention has been paid into developing Hg^{2+} sensors, which can work in the aqueous phase. Despite the preference, water-soluble Hg^{2+} sensors are still not very common. Also, the traditional chemosensors that contain sulphur moiety and have a sensing process involving the coordination of Hg^{2+} to the S atom, are not recyclable. This has incentivized the development of sustainable, recyclable sensors for the mercury ion [17, 18].

2. Types of probing agents for mercury

2.1. Chemosensors

A chemo-receptor, also referred to as chemosensor, is a sensory receptor that transfers a chemical signal into action potential. Additionally, a chemosensor detects chemical stimuli within the environment. Basically, it detects the presence of a required analyte in the solution/environment and gives a detectable signal to certify the presence of the analyte (Figure 1)

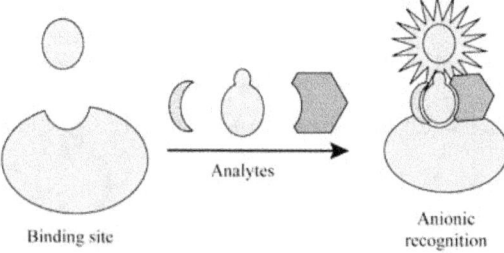

Binding site   Analytes   Anionic recognition

Figure 1: Schematic representation depicting the mechanism of action of chemosensor detecting an analyte in the solution and giving rise to a detectable signal.

Based on the type of signal generation, these chemosensors are categorized into 4 types: (1) Redox Potential, (2) Absorbance (or color), (3) Luminescence (or fluorescence) and (4) NMR relaxation times [19]. This review will focus on fluorescent chemosensors, or those which give fluorescence signals in the presence of the analyte. These fluorescent chemosensors are categorized (as depicted in Figure 2) on the basis of their synthesis, into the intrinsic, conjugate and auto-assembled varieties. The luminescent chemosensors are used as a binding site in case of intrinsic sensors or as an optical signaling unit covalently linked to a receptor unit in case of conjugated sensor. However, the basic mechanism commonly followed by the fluorescent chemosensors is primarily by photo-induced electron transfer (PET), where the donor moiety gains energy from the incoming light used for excitation of its electron(s) to the higher state, facilitating easy donation to the

7

acceptor moiety. In the due course of the process, energy released can be detected as a fluorescence signal. Internal electron transfer (ICT) is an alternate mechanism where a fraction of the electron charge is transferred between the molecular entities [20].

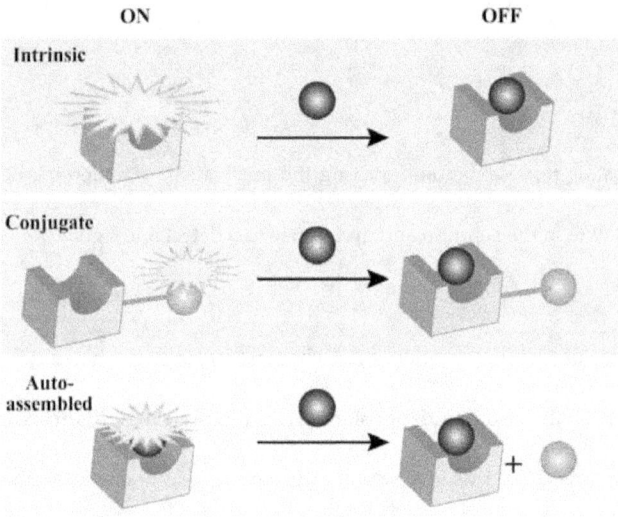

Figure 2: Types of fluorescent chemosensors based on functionality. Top panel: intrinsic; middle panel: conjugate and bottom panel: auto-assembled chemosensors. On arrow: sphere represents an "analyte".

2.2. Mercury chemosensors

Over the period of modifications being made all along on the mercury detecting probes, research has made significant progress. Starting from use of the thio-group containing organic solvents to check the black colouration generated by HgS, it goes to an extent up to the preference of metal complexing with suitable ligands and for the validation of Hg^{2+}

occurrence (Figure 3). According to the current field of exploration, most chemosensors

now being synthesized are of the fluorescent nature owing to their ease of detection, cost

effective synthesis, and easy instrumentation. The use of fluorophores in detection of this

toxic metal ion can now be seen as a common approach.

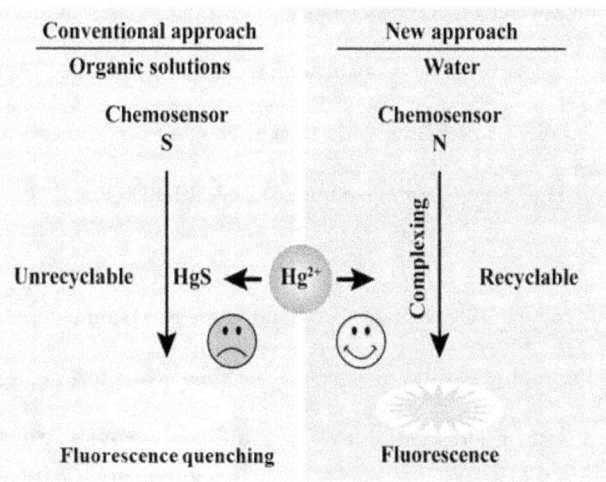

Figure 3: Common approaches using thio and nitrogen-based ligands as chemosensors

for Hg^{2+} detection. The mechanism of detection is mainly through metal complex

formation.

2.3. Types of mercury detecting chemosensors

2.3.1. Molecular Sensors

The most important strategy that needs to be understood for Hg^{2+} sensing is due to the fact that the chemical bond formation occurs using the fourth-, fifth-, and sixth-group elements as electron acceptors. During this bond formation, a strong affinity persists between electron acceptors and donors. It is the selective recognition that increases the stability and sensitivity of a sensor. This so-formed chemical bond could either be Hg–O, Hg–C, Hg–N or Hg–S bonds as described below [21].

Hg-O: These involve the cleavage of C=O bond in the rhodamine derivatives, which are excellent fluorophores used in sensors. The ring-closed form of spirolactam is colourless and is not fluorescent. But upon addition of analytes, structural change occurs and the closed spirolactam changes to the ring-opened form and results in change from colourless to an intense colouration, along with the presence of strong fluorescence. This principle has been used to design numerous mercury sensors. Hg^{2+} is known to have been able to promote the hydrolysis of isopropenyl acetate under mild conditions. In 2012, Das et al. synthesized a probe that incorporated a quinolone unit into a rhodamine 6G derivative in order to obtain different wavelengths along the fluorescence spectrum [22]. They also synthesized two rhodamine derivatives (sensor 1 and 2 in Figure 4) for recognition of mercury and copper ions. The change in the structural framework of rhodamine led to the visible change in emission intensities. The experiment was performed in CH_3CN-aqueous media and absorption was studied after reacting with Hg^{2+}. Absorption was observed around 527 nm. As the rhodamine derivative was made to interact with Hg^{2+}, the luminescence intensity observed at 557 nm is enhanced with a detection limit of 0.35 ppb for Hg^{2+}. The ratiometric sensor enables the interaction of rhodamine with 1,8-naphthalimide derivatives as two fluorophores for the detection of Hg^{2+}. The non-radiative transfer of excitation energy is induced due to the binding of Hg^{2+} from the donor naphthalimide to the acceptor xanthene moiety.

Figure 4: Synthesis of rhodamine derivatives (1 and 2) as the sensors for the detection of Hg^{2+} and Cu^{2+}. (i) o-phenylenediamine, N,N′-icyclohexylcarbodiimide (DCC), 4-Dimethylaminopyridine (DMAP), stirred in CH_2Cl_2 for 19 h; (ii) AcOH, heat; (iii) 50% trifluoroacetic acid (TFA) in CH_2Cl_2, stirred for 3 h. Reproduced with permission from (Prasenjit et al. 2014) by Royal Society of Chemistry.

The ability of the sensor to detect the presence of different divalent ions using the UV-visible and fluorescence spectroscopy tool is shown below (Figure 5). The change in the emission spectra occurred at 336 nm when 15 equivalents of the metal ions were excited in CH_3CN containing 5% DMSO. It was observed that clearly the sensor had strong interaction with Hg^{2+} and a little interaction with Zn^{2+} as well. Fluorescence data confirmed emission at 336 nm, which became clearly visible when $Hg(ClO_4)_2$ was added through titration and a red shift of 5 nm was observed as well. This red shift was caused due to the stabilization of the fluorophore in its excited state compared to the ground state on binding of the cation (Figure 6).

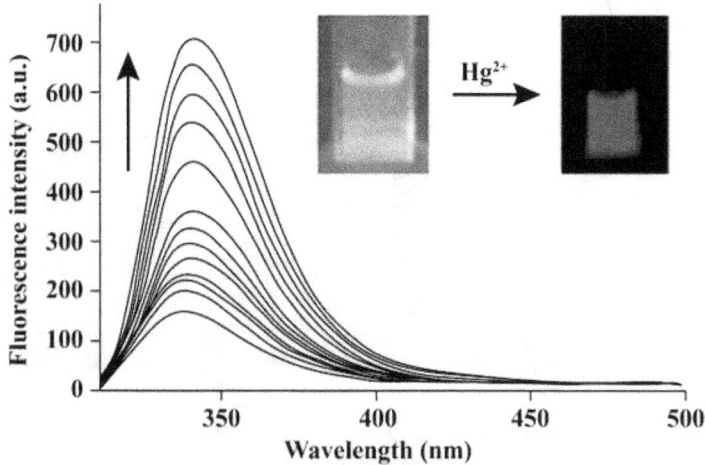

Figure 5: Graph manifesting the change in emission of ligand **1** upon gradual addition of

Hg^{2+} ion at 336 nm. Inset shows the switch-on fluorescence upon interaction of

chemosensor with Hg^{2+} ion.

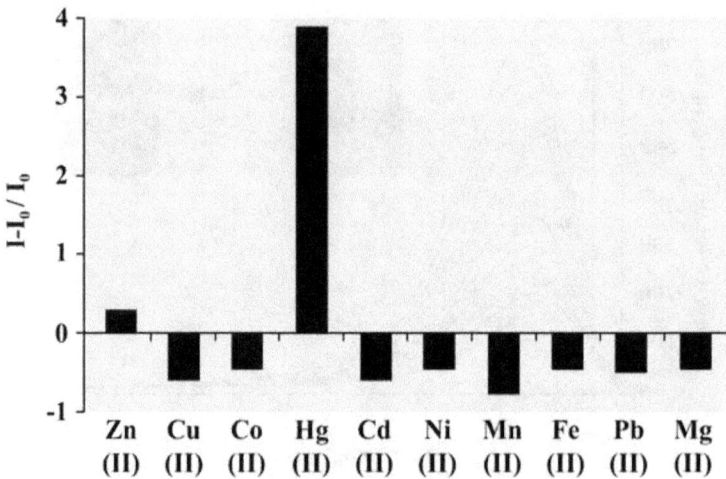

Figure 6: Change in fluorescence ratio of ligand **1** upon addition of 15 equivalents of cations at 336 nm. This red shift was caused due to the stabilization of the fluorophore in its excited state.

Hg-S: Formation of Hg-S occurs by C=S or C-S-C bond cleavage. Chemical properties of sulfur and oxygen are somewhat similar because they belong to the same group on the periodic table. As discussed above, extensive studies were performed for Hg^{2+} detection by Hg-O bond formation and the cleavage of C=O bond. Hence, Figure 7 depicts the synthesis of chemosensors with the Hg-S bond formation and its role in the detection of Hg^{2+} ion.

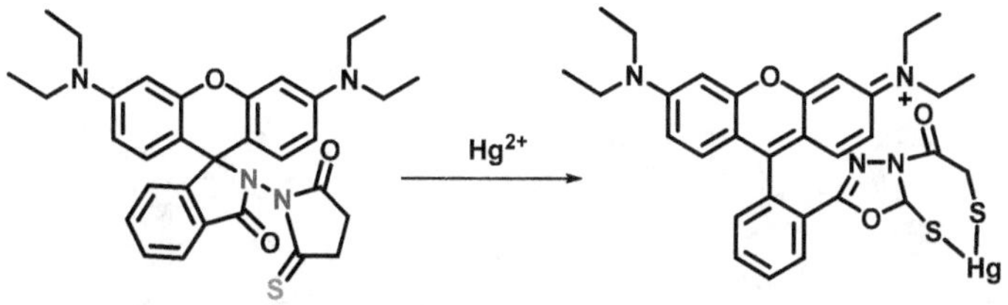

Figure 7: Proposed sensing mechanism of Hg^{2+} ion through Hg-S bond formation from fluorescein-based thiosemicarbazide group.

Yang's group designed and synthesized a fluorescein-based chemodosimeter for Hg^{2+} detection [23]. In addition to this probe (I) to Hg^{2+} in aqueous methanol solution, the thiosemicarbazide group undergoes a desulfurization reaction that involves irreversible cleavage of C=S bond and forms the corresponding oxidiazole (Fig. 8). Herein the Hg^{2+} become stable by the removed sulfur through Hg-S bond formation. Hence, the visible colouration and fluorescence emission caused by the oxadiazole states that the maximum linear response of the sensor to Hg^{2+} is 0.8 μM, with a detection limitation of 9.4 nM in methanol-H$_2$O. Changing the substitution with a multi nitro moiety made this probe a reversible sensor (II), which is highly fluorescent (Figure. 8).

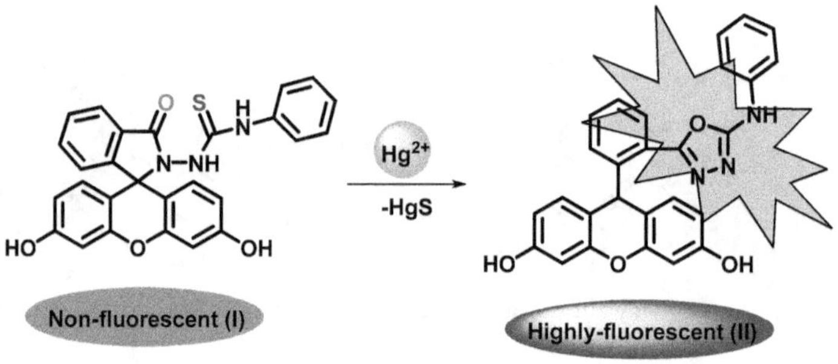

Probe (I)

Figure 8: Desulphurization of **I** by Hg²⁺ ion leading to the highly fluorescent oxadizole **(II).** Reproduced with permission from (Yang et al. 2007) by Elsevier.

Guiqui et al. reported the design and synthesis of Schiff base with rhodamine B thiohydrazide and a quinoline moieties [24]. When probe I was made to react with Hg²⁺, a clear pink color turn-on appears, which causes 106-fold fluorescence enhancement with an absorption maximum at 565 nm. This probe had a response range of 0.1 to 10 µM and detection limit of 8.5 nM. Similarly, a fluorescent chemosensor (Figure 9) with a receptor composed of S atom and an alkene moiety when reacted with Hg²⁺ metal ion showed a large fluorescent enhancement response (approximately 1000-fold). There was maximum absorption at 561 nm with high selectivity and sensitivity, while having an LOD of 27.5 nM [25].

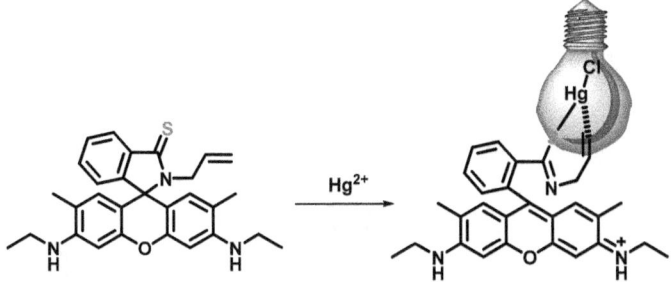

Figure 9: Fluorescent chemosensor with S and alkene moiety showing turn-on response upon addition of Hg^{2+} ion. Reproduced with permission from (Lin et al. 2010) by Royal Society of Chemistry.

Hg-N: Hancock's research group found a "turn-off" fluorescent sensor for Hg^{2+} ions in solution [26]. They basically explored through the tendency of Hg^{2+} ions to quench the fluorescence of a potential fluorescent sensor. Hancock's and group worked with multiple divalent ions and studied the chelation-enhanced fluorescent (CHEF) effect [27] and observed that Hg^{2+} showed way lesser chelation effect in comparison to Zn^{2+} and Cd^{2+}. This reflects the heavy atom effect, which could have been caused by increasing spin-orbit coupling constants (Figure 10). Unlike the other divalent ions, mercury is usually good at quenching the fluorescence, hence causing a negative "turn-off" chelation enhanced quenching (CHEQ) effect [28]. They worked with coordinated and tethered fluorophores such as *N,N*-bis(2-methylquinoline)-2-(2-aminomethyl)pyridine (DQPMA) and (N-(9-anthracenylmethyl)-N,N-di-(picolyl)amine) (ADPA).

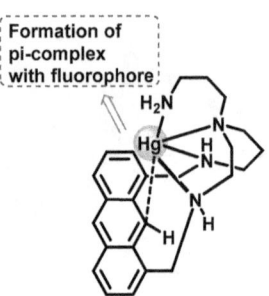

Figure 10: Possible Structure of a Mercury (II) Complex of the ligand that leads to Quenching of the Fluorescence via formation of a π Complex between Hg^{2+} and the Anthracenyl Fluorophore. Reproduced with permission from (Wei Gan et al. 2005) by Elsevier.

As manifested in Figure 11, Lee et al showed that Hg^{2+} contained in a solution prominently quenches the fluorescence of ADPA, which was investigated using crystallography for Hg-ADPA complexes, and the conclusion was supported by DFT calculations [29]. It was also found that coordination of halide ions or hydroxide to the Hg^{2+}-ADPA complex could displace the anthracenyl group coordinated to Hg^{2+} and restore its fluorescence, hence forming a novel type anion sensor. The addition of more covalent binding ligands to the mercury ion tends to lengthen Hg-N bonds to the saturated N-donor atom of the dipicolylamine part of ADPA. The information was supported by crystallography and DFT calculations. Moreover, DFT calculations

18

also show that the quenching effect observed in Hg^{2+}-ADPA complex is the cause of π-interaction of mercury ion with the fluorophore, leading to the decrease of the electronic transition probability to ground state from excited state. It was concluded that for designing a turn-on sensor for mercury ion, it's important to have a covalently bonded ion such as S-donor atom for chelating.

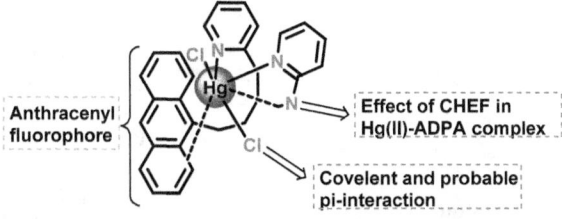

Figure 11: Formation of stable Hg^{2+}/ADPA complex which helps in restoring its fluorescence. Reproduced with permission from (Lee et al. 2012) by American Chemical Society.

Hg interaction with both N & S moiety

Lippolis' group reported that the 1,10-phenanthroline unit was an integral part of the macrocyclic structure, which include prominent thioether donors and were potential fluorescent chemosensors for soft metal sensors. These phenanthroline derivatives were then used to study the toxicity of divalent ions of lead, cadmium, and mercury [30].

Formation of Hg-Se bond

This occurs by the cleavage of the C-Se-C bond. Being of the same group as sulphur, selenium does show strong affinity to Hg^{2+} similar to S [31].

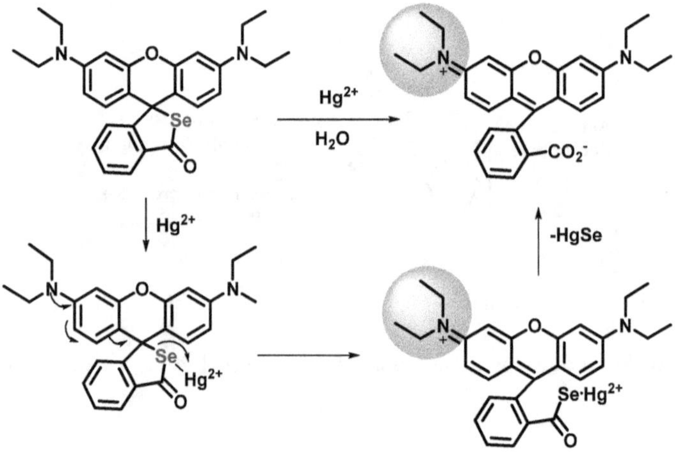

Figure 12: Sensing mechanism of rhodamine-based chemosensor for Hg^{2+} detection.

It was observed that the selonacton based sensor exhibited intensive fluorescence enhancement at 580 nm over the test range of concentration from 0 to 30 nM. Moreover, this sensor showed an observable optical response to methylmercury species in HEPES buffer as well as cells/zebrafish. The unusual fluorescence enhancement is induced due to the formation of Hg-Se bond followed by eventual deselenation reaction (Figure 12).

Table 1 shows some of the newly developed biosensing techniques.

Probe type	Probing agents	LOD (in ppb)	References
Optical-based biosensor	Mercaptopropionic acid-homocysteine-PDCA-modified gold nanoparticles	5	32
Electrochemical-based senso	Protein-functionalized reduced graphene oxide (rGO)	0.2	33
Paper-based chemosensor	Sulfur bearing phenanthridines	4.1	34

2.3.2. Formation of Hg-C bond

The need for this kind of sensors occurred because it was observed that Hg-S binding had drawbacks, like the oxidation of sulfide over long-term storage. Hence, binding of Hg^{2+} using the Hg-C bond could be considered a fair alternative. As reported by Koide's group, an oxidation-resistant fluorescent sensor for mercury was based on alkyne oxymercuration mechanism [35]. Results from this study suggested that the binding of Hg^{2+} to dissolved organic matter (DOM; hydrophobic acids isolated from the Florida Everglades by XAD-8 resin) under ambient conditions (with very low Hg/DOM ratios) is controlled by a very small number of DOM molecules containing a highly reactive thiol functional group (Figure 13). Thus, the distribution coefficients of Hg/DOM that were used for studying the biogeochemical behavior of Hg in natural systems need to be determined at low Hg/DOM ratios. With the addition of Hg^{2+} to this sensor system, hydration of alkynes occurs to form ketones.

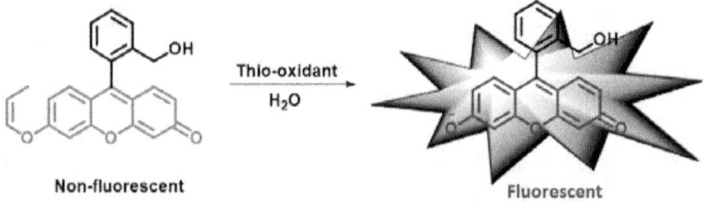

Non-fluorescent **Fluorescent**

Figure 13: Non-fluorescent DOM upon reacting with a thio-based mercury compound in the presence of an oxidant becomes a highly fluorescent system. Reproduced with permission from (Ando et al. 2011) by American Chemical Society.

3. Hg²⁺ detection via amide binding site

It is well-known that amide contains nitrogen and oxygen and that mercury (II) ions have a profound biding affinity towards amides. For instance, rhodamine B phenyl hydrazide selectively recognize and coordinate with Hg²⁺ [36]. Similarly, Sung-Kyun *et.al,* synthesized a rhodamine based chemosensor to detect mercury ions in zebrafish tissue and organs [37]. This chemosensor responded rapidly and readily to mercury ions in aqueous media at room temperature. Real time monitoring showed that the mercury ion uptake by cells reaches the saturation limit in roughly 20-30 minutes. They uncovered a rhodamine based chemosensor which unlike any other could react irreversibly with mercury ions (Figure 14).

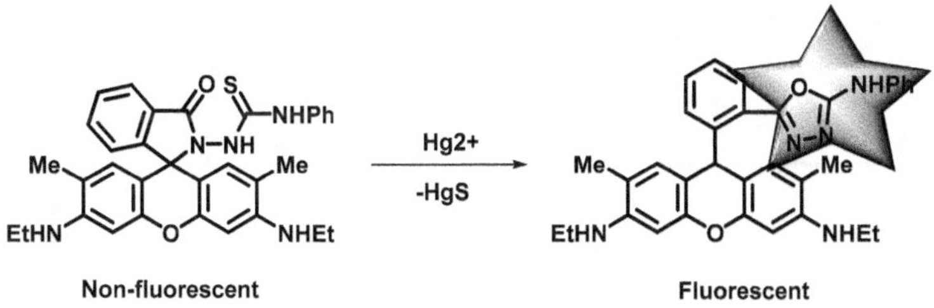

Figure 14: Conversion of non-fluorescent to strongly fluorescent rhodamine-based system by mercury ions. Reproduced with permission from (Ko et al. 2006) by American Chemical Society.

4. Hg²⁺ detection through sulphonamide binding site

Cao et al., developed a new pyrenyl-appended triazole system (Figure 15) for fluorescent recognition in acetonitrile aqueous solution [38]. It is also a selective and sensitive chemosensor. In the presence of mercury ion in the environment and in acetonitrile aqueous solution, the fluorescence intensity of the complex decreased by 80%. Complex formation with mercury (II) salt was observed with a distinctive fluorescent selectivity for Hg²⁺ ion.

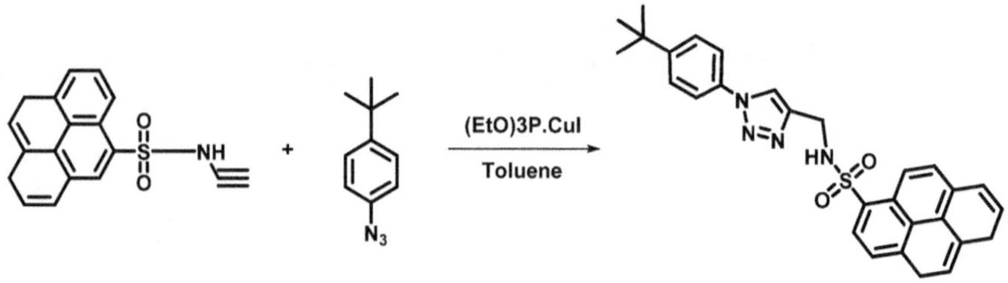

Figure 15: Synthetic route for obtaining a novel pyrenyl-appended triazole system for the detection of Hg²⁺ ions. Reproduced with permission from (Cao et al. 2013) by Elsevier.

5. Schiff base ligands in Hg²⁺ ion detection

Wei *et.al.* in 2014 reported a double naphthalene-based Schiff base with mercury detecting properties [39]. Authors developed a fluorescent chemosensor that was cheap, sensitive, and highly selective to mercury ions. The imine present in the ligand could be oxidized to amide with DMSO, and this could confer the coordination capacity required for coordination with Hg²⁺. Later spectroscopic analysis was performed in order to study

the sensitivity and selectivity of the sensor. The senor was synthesized by a simple and

low-cost Schiff base reaction of α-napthaldehyde and α-napthylamine, with a catalytic

amount of acetic acid in hot absolute ethanol for 4 h. The proposed mechanism is shown

in Figure 16.

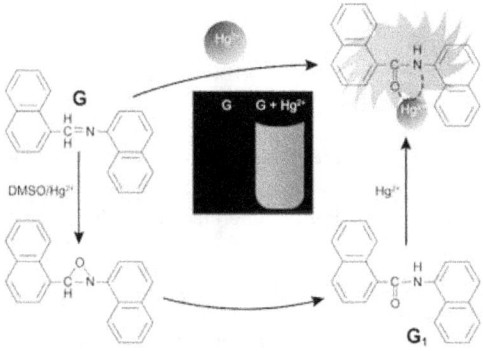

Figure 16: Double-naphthalene Schiff base chemosensor (G) towards Hg²⁺ probing in

DMSO solution.

The probable mechanism of detection can be explained as follows. The presence of DMSO

and Hg²⁺ led to the formation of O-Hg²⁺-N bond, which eventually got converted into a

C=N of the ligand and was oxidized to O=C-NH of the sensor(G₁). A distinct blue

colouration was visible under 365 nm UV lamp on allowing the Schiff base to form complex with mercury ions. The ligand was found to be highly selective due to mercury divalent ions and did not bind with any other cation, even after 10-fold increase in cation concentration. The LOD was recorded to be 5.595×10^{-8} M, hence the sensor could serve as a fluorescent sensor for Hg^{2+} ion.

6. With porphyrine binding group

Li *et.al.* unveiled a novel napthalimide-porphyrin hybrid based fluorescent probe that could detect the presence of Hg^{2+} ratiometrically in aqueous solution and living cells [40]. It is designed by keeping two independent Hg-sensors at their maximum excitation wavelengths. The ultimate probe hence generated has a limit of detection, 2.0×10^{-8} M. Moreover, the ratiometric fluorescence change of the sensor changes significantly and is highly specific for mercury ions, even in the presence of cellular metal ions such as Na^+, K^+, Mg^{2+}, or Ca^{2+}. The results are similar in the presence of essential metal ions like Zn^{2+}, Fe^{3+}, Fe^{2+}, Cu^{2+}, Mn^{2+}, Co^{2+}, and Ni^{2+}, or heavy metal ions like Ag^+, Pb^{2+}, Cr^{3+}, and Cd^{2+}, which enable the selective requirements for environmental and biomedical monitoring application. The recovery of Hg^{2+} from water samples collected from any random location prove the feasibility of the above designed probe and justifies its practicality. Because of its ratiometric imaging of the Hg^{2+} in cells with enough resolution, it indicates its efficiency as a novel sensor. Mostly on-off probes are reported to be favourable for bioimaging applications. Most fluorescent probes based on single emission intensity changes are usually affected by a variety of factors, like efficiency of the instrumentation, concentration of sensor molecule, stability under photo-illumination, and the microenvironment around the sensor molecule. These drawbacks can be overcome by using ratiometric probes that are not prone to such issues. On the other hand, they involve observation of changes in the ratio of intensities of emission at two wavelengths when adding the target, which is beneficial for increasing the variable range and provides alarm for environmental hazards.

7. Nanostructures as efficient Hg^{2+} probes

In recent years, extending the chemosensor research to a direction with a constructive approach is setting in, particularly 1D assemblies of nanoparticles i.e. chain of attached molecules can help understand the vivid range of phenomenon, starting from processes in living organisms' bodies to quantum mechanism of nanometre-scale systems. The sensing mechanism of AuNPs is originated from the aggregation of in the presence of Hg^{2+} [41-49].

Two main reasons that support the existence of binding mode is (i) the negative superficial charge of the silver nanoparticle, and (ii) the ionic radium of this metal ion is small enough to favour the accommodation of metal ion in the inner cavity. A similar effect was also observed in the recently reported systems. The chelation interaction between Hg^{2+} ion and the carboxylate groups of chemosensor L located on the surface of AuNPs and AgNPs is responsible for the selective formation

of chains between the NPs modulated for Hg²⁺ ions (Figure. 17) [50-55].

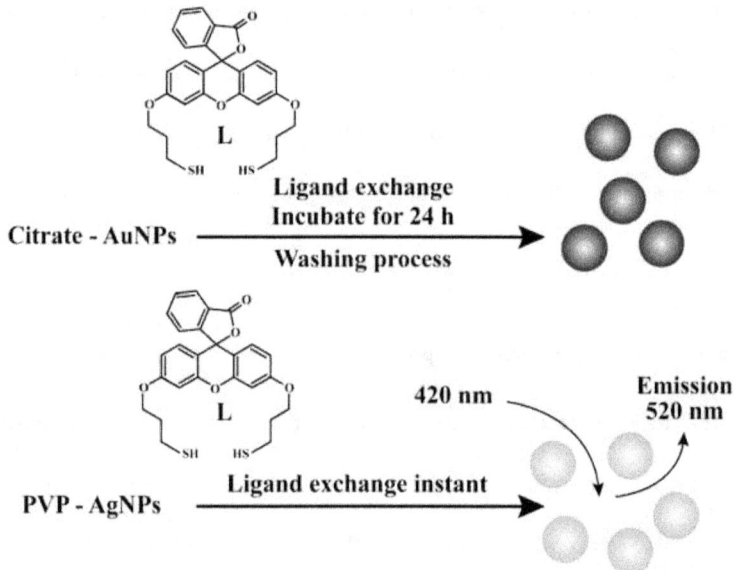

Figure 17: The surface coated of AuNPs and AgNPs using Ligand (L) and their use

in probing of Hg²⁺ ion.

8. Conclusions

Because of the dedicated efforts by researchers over many years, multiple chemo-sensors

have been developed and a pathway towards controlling the toxicity could be

constructed. Most traditional chemo-sensors include a sulphur moiety involving

mechanism that is driven by mercury's affinity for thio-groups, hence yielding Hg-S formation, ring opening of spirocyclic systems (rhodamine and fluorescein, etc.), conversion of thiocarbonyl compounds into their carbonyl analogues, or a sequential desulfurization reaction. This thiophilic approach isn't completely reliable. Hence, formation of metal complexes with fluorophores is a better approach. Furthermore, the "heavy atom" effect for Hg(II) favors an enhanced spin-orbit coupling constant (ζ) and induces a strong luminescence quenching of the bound luminophore. Moreover, the high hydration enthalpy (1824 kcal mol-1) for Hg^{2+} ion is another challenging fact for the scientists to achieve the seemingly simple but tricky issue of Hg^{2+} recognition with luminescence on or enhancement response, either in aqueous environment or in physiological conditions. Luminescence enhancement is crucial for designing the reagent for imaging application and detection of the cellular uptake of this ion. To sum up, a brief but extensive literature survey is carried out to account on the recent developments on the design and synthesis of chemical sensors along with the reports on colorimetric reagent that are conducive for infiled sample analysis with yes-no type binary response.

Acknowledgements

The efforts of the researchers working in the field of developing chemical sensors for mercury detection are highly appreciated.

Conflict of interest

Authors declare no conflict of interests.

References

1. Bharath Kumar Momidia, Venkatadri Tekuria and Darshak R. Trivedi. Inorganic Chemistry Communications, 2016, 74, 1-5.

2. Bernard Valeur, Isabelle Leray. Coordination Chemistry Reviews, 2000, 205, 3-40.

3. Yu Y, Lin LR, Yang KB, Zhong X, Huang RB, Zheng LS. Talanta, 2006, 69, 103-106.

4. Amin-Zaki L., Elhassani S., Majeed M.A., Clarkson T.W.,Doherty, R. A.; Greenwood, M. Pediatrics, 1974, 54, 587–595.

5. Liu D.B, Qu W.S, Chen W.W, Zhang W, Wang Z, Jiang X, Y. Anal. Chem. 2010, 82, 9606–9610.

6. Xu Y, Deng L, Wang H, Ouyang X, Zheng J, Li J, Yang R. Chemical Communications, 2011, 47, 6039-6041.

7. Wu J, Li L, Zhu D, He P, Fang Y, Cheng G. Anal Chim Acta, 2011, 694, 115-119.

8. Jung-Duck Park and Wei Zheng. J Prev Med Public Health. 2012, 45, 344–352.

9. Björkman L, Lundekvam BF, Laegreid T, Bertelsen BI, Morild I, Lilleng P. Environ Health, 2007, 6, 30.

10. Clarkson TW, Magos L. Crit Rev Toxicol., 2006, 36, 609–662.

11. Cariccio VL, Samà A, Bramanti P, Mazzon E. Biol Trace Elem Res., 2019, 187, 341-356.

12. Kevin M. Rice, Ernest M. Walker, Jr, Miaozong Wu, Chris Gillette, and Eric R. Blough. J Prev Med Public Health, 2014, 47, 74–83.

13. Byeong-Jin Ye, Byoung-Gwon Kim, Man-Joong Jeon, Se-Yeong Kim, Hawn-Cheol Kim, Tae-Won Jang, Hong-Jae Chae, Won-Jun Choi, Mi-Na Ha, and Young-Seoub Hong. Ann Occup Environ Med. 2016, 28, 5.

14. Ying Gao, Zeming Shi, Zhou Long, Peng Wu, Chengbin Zheng, Xiandeng Hou. Microchemical Journal, 2012, 103, 1-14.

15. Tomasz Wasilewski, Jacek Gębicki, Wojciech Kamysz. TrAc Trends in Analytical Chemistry, 2021, 142, 116330.

16. Qisong Zhang, Jian Zhang, Hujin Zuo, Chengyun Wang, Yongjia Shen. Tetrahedron, 2017, 73, 2824-2830.

17. Mahnaz D. Gholami, Sergei Manzhos, Prashant Sonar, Godwin A. Ayoko, Emad L. Izake. Analyst, 2019, 144, 4908-4916.

18. Wen-juan Qu, Guo-ying Gao, Bing-bing Shi, Tai-bao Wei, You-ming Zhang, Hong Yao. Sensors and Actuators B: Chemical, 2014, 204, 368-374.

19. Hongmin Jia, Ming Yang, Qingtao Meng, Guangjie He, Yue Wang, Zhizhi Hu, Run Zhang, and Zhiqiang Zhang. Sensors, 2016, 16(1), 79.

20. Kyle P. Carter, Alexandra M. Young, Amy E. Palmer, Fluorescent Sensors for Measuring Metal Ions in Living Systems. Chemical Reviews, 2014, 114, 8, 4564-4601.

21. Bruno B. Campos, Manuel Algarra, Beatriz Alonso, Carmen M. Casado and Joaquim C. G. Esteves da Silva. RSC Advances, 2009, 134, 2447–2452.

22. Prasenjit Mahato, Sukdeb Saha, Priyadip Das, Hridesh Agarwalla and Amitava Das. RSC Advances, 2014, **4**, 36140-36174.

23. Yang XF, Li Y, Bai Q. Analytica Chimica Acta, 2006, 584(1):95-100.

24. Guiqiu Chen, Zhi Guo, Guangming Zeng and Lin Tang. Analyst, 2015, 140, 5400-5443.

25. Lin, W. Y.; Cao, X. W.; Ding, Y. D.; Yuan, L.; Yu, Q. X. Organic and Biomolecular Chemistry, **2010**, 8, 3618 3620.

26. Wei Gan, S. Bart Jones, Joseph H. Reibenspies, Robert D. Hancock. Inorganica Chimica Acta, **2005**, *358* (13), 3958-3966.

27. Michael E. Huston, Engin U. Akkaya, and Anthony W. Czarnik. The Journal of American Chemical Society. **1989**, 111, 8735-8737.

28. Engin U. Akkaya, Michael E. Huston, and Anthony W. Czarnik. The Journal of American Chemical Society. **1990**, 112, 3590-3593.

29. Hyunjung Lee, Hee-Seung Lee, Joseph H. Reibenspies and Robert D. Hancock. Inorganic Chemistry, **2012**, 51, 10904–10915.

30. Andrea Bencini and Vito Lippolis. Coordination Chemistry Reviews, **2010**, 254, 2096-2180.

31. Parthiban Venkatesan, Natesan Thirumalivasan, Shu-Pao Wu. A rhodamine-based chemosensor with diphenylselenium for highly selective fluorescence turn-on detection of Hg^{2+} *in vitro* and *in vivo*. RSC Advances, **2017, 7**, 21733-21739.

32. Cesar S. Huertas, Olalla Calvo-Lozano, Arnan Mitchell and Laura M. Lechuga. Advanced Evanescent-Wave Optical Biosensors for the Detection of Nucleic Acids: An Analytic Perspective. Frontiers in Chemistry,2019, 7, 724.

33. Sudibya HG, He Q, Zhang H, Chen P (2011) Electrical detection of metal ions using field-effect transistors based on micropatterned reduced graphene oxide films. ACS Nano 5: 1990-1994.

34. Marimuthu Ponram, Umamahesh Balijapalli, Baskaran Sambath, Sathiyanarayanan Kulathu Iyer, Venkatachalapathy B, Ravichandran Cingaram and Karthikeyan Natesan Sundaramurthy. (2018) Development of paper-based chemosensor for the detection of mercury ions using mono- and tetra-sulfur bearing phenanthridines. New Journal of Chemistry, 2018, 42, 8530-8536.

35. Ando, S.; Koide, K. Development and Applications of Fluorogenic Probes for Mercury(II) Based on Vinyl Ether Oxymercuration. The Journal American Chemical Society, 2011, 133, 2556–2566.

36. Krishnendu Pramanik, Priyabrata Sarkar, Dipankar Bhattacharyay. Semi-quantitative colorimetric and supersensitive electrochemical sensors for mercury using rhodamine b hydrazide thio derivative. Journal of Molecular Liquids. 2019, 276, 141-152.

37. Sung-Kyun Ko, Young-Keun Yang, Jinsung Tae, and Injae Shin, *In Vivo* Monitoring of Mercury Ions Using a Rhodamine-Based Molecular Probe. The Journal America Chemical Society, 2006, 128, 14150-14155.

38. Qian-Yong Cao, Yuan-Ming Han, Hong-Ming Wang and Yu Xie. A new pyrenyl-appended triazole for fluorescent recognition of Hg^{2+} ion in aqueous solution. Dyes and Pigments, 2013, 99, 798-802.

39. Tai-bao Wei, Guo-ying Gao, Wen-juan Qu, Bing-bing Shi, Qi Lin, Hong Yao, You-ming Zhang. Selective fluorescent sensor for mercury(II) ion based on an easy to prepare double naphthalene Schiff base. Sensors and Actuators B: Chemical, 2014, 199, 142-147.

40. Chun-Yan Li, Xiao-Bing Zhang, Li Qiao, Yan Zhao, Chun-Mei He, Shuang-Yan Huan, Li-Min Lu, Li-Xin Jian, Guo-Li Shen, and Ru-Qin Yu. Naphthalimide–Porphyrin Hybrid Based Ratiometric Bioimaging Probe for Hg^{2+}: Well-Resolved Emission Spectra and Unique Specificity. Analytical Chemistry, 2009, 81, 9993–10001.

41. Alivisatos P, Nature Biotechnology. 2004, 22, 47–52.

42. Xi Zeng, Lei Dong, Chong Wu, Lan Mu, Sai-Feng Xue, Zhu Tao, Highly sensitive chemosensor for Cu(II) and Hg(II) based on the tripodal rhodamine receptor, Sensors and Actuators B: Chemical, 141, 2, 2009, 506-510.

43. Da-Hye Kim, Junho Seong, Hyunsook Lee, Keun-Hyeung Lee, Ratiometric fluorescence detection of Hg(II) in aqueous solutions at physiological pH and live cells with a chemosensor based on tyrosine, Sensors and Actuators B: Chemical, 2014, 196, 421-428.

44. Nuriman, Bambang Kuswandi, Willem Verboom, Selective chemosensor for Hg(II) ions based on tris[2-(4-phenyldiazenyl)phenylaminoethoxy]cyclotriveratrylene in aqueous samples, Analytica Chimica Acta, 2009, 655, 75-79.

45. Zhuo Wang, Deqing Zhang, Daoben Zhu, A sensitive and selective "turn on" fluorescent chemosensor for Hg(II) ion based on a new pyrene–thymine dyad, Analytica Chimica Acta, 2005, 549, 10-13.

46. Hossein Abdolmohammad-Zadeh, Elaheh Rahimpour, A novel chemosensor based on graphitic carbon nitride quantum dots and potassium ferricyanide chemiluminescence system for Hg(II) ion detection, Sensors and Actuators B: Chemical, 2016, 225, 258-266.

47. Xiaohong Peng, Yujiao Wang, Xiaoliang Tang, Weisheng Liu, Functionalized magnetic core–shell Fe3O4@SiO2 nanoparticles as selectivity-enhanced chemosensor for Hg(II), Dyes and Pigments, 2011, 91, 26-32.

48. Nai-bo Zhang, Jian-jun Xu, Chen-guang Xue, Core–shell structured mesoporous silica nanoparticles equipped with pyrene-based chemosensor: Synthesis, characterization, and sensing activity towards Hg(II), Journal of Luminescence, 2011, 131, 2021-2025.

49. Venkatesan Muthukumar, Sathiyanarayanan KulathuIyer, Twisted pyrene with perfect hetero atomic cavity optical sensor for Hg22+ and Pb2+, Inorganic Chemistry Communications, 2020, 121, 108187.

50. Nguyen Khoa Hien, Nguyen Chi Bao, Nguyen Thi Ai Nhung, Nguyen Tien Trung, Pham Cam Nam, Tran Duong, Jong Seung Kim, Duong Tuan Quang, A highly sensitive fluorescent chemosensor for simultaneous determination of Ag(I), Hg(II), and Cu(II) ions: Design, synthesis, characterization and application, Dyes and Pigments, 2015, 116, 89-96.

51. Juyoung Yoon, Norman E. Ohler, David H. Vance, Wade D. Aumiller, Anthony W. Czarnik, A fluorescent chemosensor signalling only Hg(II) and Cu(II) in water, Tetrahedron Letters, 1997, 38, 3845-3848.

52. Ali Coskun, M. Deniz Yilmaz, and Engin U. Akkaya Bis(2-pyridyl)-Substituted Boratriazaindacene as an NIR-Emitting Chemosensor for Hg(II), Organic Letters, 2007, 9, 607–609.

53. Zhi-Xiang Han, Hong-Yuan Luo, Xiao-Bing Zhang, Rong-Mei Kong, Guo-Li Shen, Ru-Qin Yu, A ratiometric chemosensor for fluorescent determination of Hg2+ based on a new porphyrin-quinoline dyad, Spectrochimica Acta Part A: Molecular and Biomolecular Spectroscopy, 2009, 72, 1084-1088.

54. Reham Ali, Ibrahim A.I. Ali, Sabri Messaoudi, Fahad M. Alminderej, Sayed M. Saleh, An effective optical chemosensor film for selective detection of mercury ions, Journal of Molecular Liquids, 2021, 336, 116122.

55. Buddhadeb Sem, Manjira Mukherjee, Siddhartha Pal, Koushik Dhara, Sushil Kumar Mandal, Anisur Rahman Khuda-Bukhsh and Pabitra Chattopadhyay, A water soluble FRET-based ratiometric chemosensor for Hg(II) and S2− applicable in living cell staining , RSC Advaces, 2014, 4, 14919-14927.

Publisher: Eliva Press SRL

Email: info@elivapress.com

Eliva Press is an independent publishing house established for the publication and dissemination of academic works all over the world. Company provides high quality and professional service for all of our authors.

Our Services:
Free of charge, open-minded, eco-friendly, innovational.

-Free standard publishing services (manuscript review, step-by-step book preparation, publication, distribution, and marketing).
-No financial risk. The author is not obliged to pay any hidden fees for publication.
-Editors. Dedicated editors will assist step by step through the projects.
-Money paid to the author for every book sold. Up to 50% royalties guaranteed.
-ISBN (International Standard Book Number). We assign a unique ISBN to every Eliva Press book.
-Digital archive storage. Books will be available online for a long time. We don't need to have a stock of our titles. No unsold copies. Eliva Press uses environment friendly print on demand technology that limits the needs of publishing business. We care about environment and share these principles with our customers.
-Cover design. Cover art is designed by a professional designer.
-Worldwide distribution. We continue expanding our distribution channels to make sure that all readers have access to our books.

www.elivapress.com

www.ingramcontent.com/pod-product-compliance
Lightning Source LLC
Chambersburg PA
CBHW051300170526
45165CB00004B/1785